Science Activity Books

GEOLOGY

Created by The Good and the Beautiful Team

Cover illustrated by Sandra Eide

Pages illustrated by Amanda Gulliver & Shannon Vogus

Cover design by Phillip Colhouer

Trace the names of the landscapes and color the pictures.

forest

desert

beach

mountains

Color the layers of Earth. Color sections with a 1 green. Color sections with a 2 blue. Color sections with a 3 red. Parent tip: Color the images of the crayons with the correct colors for the child to refer to.

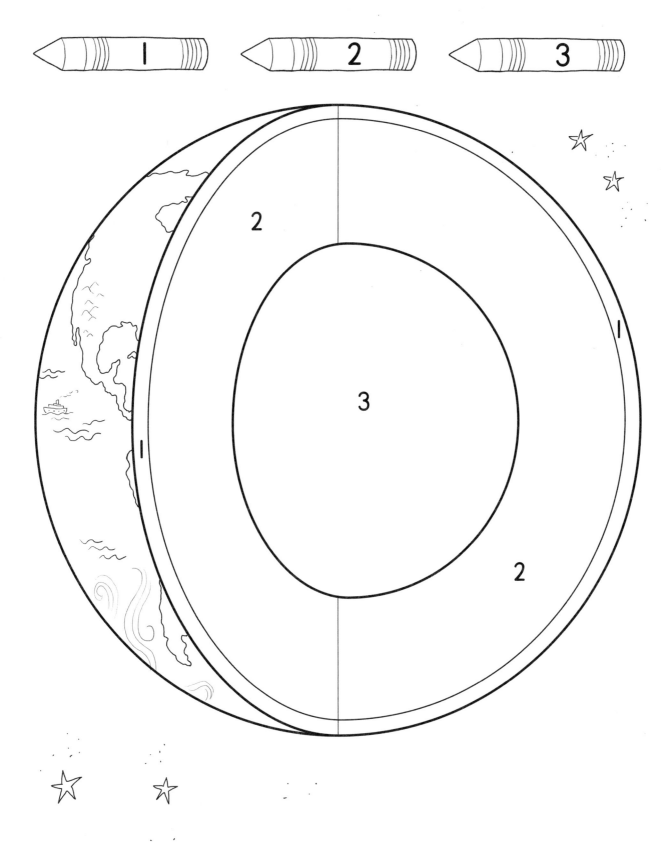

Lesson 2

Turn this page a quarter of a turn to the left.
Trace the dotted lines that make up the world's
tectonic plates. Color the picture.

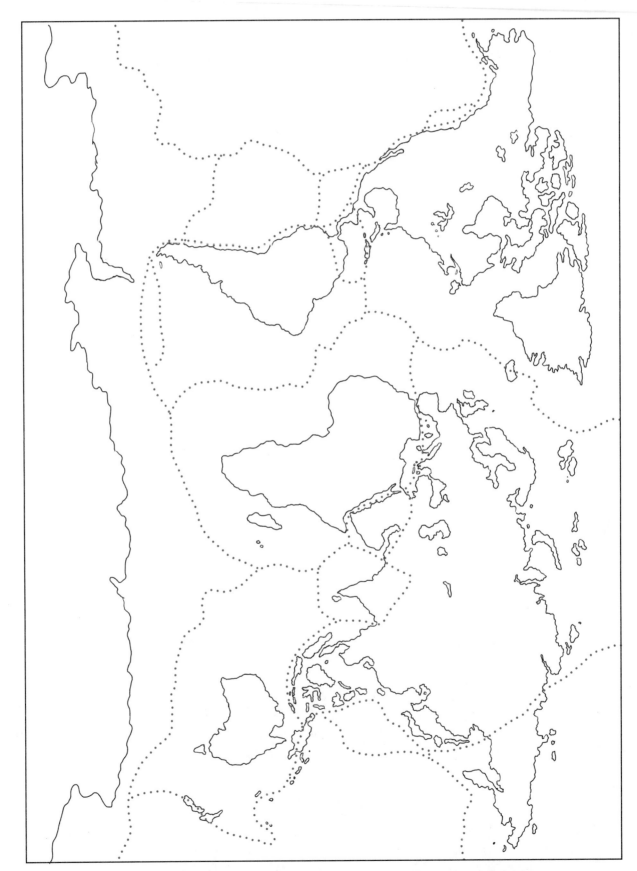

Circle the six items in the top picture that you do not see in the bottom picture. Color the top picture.

Trace the words that describe the steps you would take to protect yourself during an earthquake. Color the picture.

Draw lines to connect the dots, starting at the number 1. Color the picture. Note: There are two separate sets of numbers.

Draw lines to connect the matching volcanoes. Color the volcanoes.

Using a red crayon, trace the path of the magma at the bottom of the page up through earth and out the tops of each volcano. Color the pictures.

Lesson 5

Cross out the six silly things that do not belong in the picture of the volcano. Color the picture.

Draw lava coming up and out of the top of the volcano and side vents. Add a cloud of ash and smoke above the volcano. Color the picture.

Lesson 6

Look at the pattern in each row. Draw the rock that comes next in the blank space to the right. Color the rocks you drew.

Color sections of the hot spring with a 1 red. Color sections with a 2 yellow. Color sections with a 3 blue. Color sections with a 4 brown. Color sections with a 5 green. Parent tip: Color the images of the crayons with the correct colors for the child to refer to.

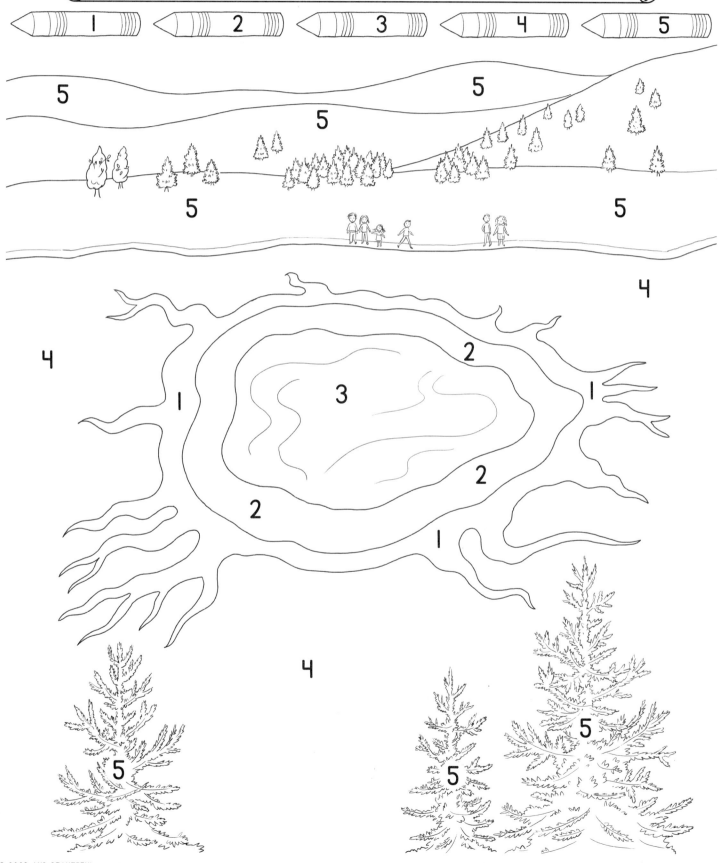

Cross out the pictures in the key as you find them in the museum. Color the picture.

Color the two crystals in each row that are the same size.

Draw lines to connect the matching rocks, gems, and crystals. Color the pictures.

Cross out the six silly things that do not belong in the picture. Color the picture.

Lesson 9

Color the pictures at the bottom of the page. Cut out the pictures and tape or paste each picture into the correct box.

Rocks	Not Rocks

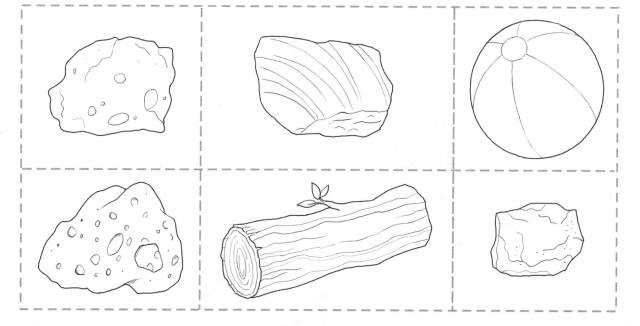

Begin at the start and trace your way through the maze to help the hikers get to Devil's Tower. Color the picture.

finish

start

Cross out the six silly things that do not belong in the picture of the cliff. Color the picture.

Find and color only the rocks with layers.

Lesson 11

Trace each line from the rocks to see how they change form. Color the rocks.

Look at the pictures in the graph at the bottom of the page. Color one box next to each item every time you find it in the picture.

Lesson 12

Add five diamonds, eight crystals, and six rocks on the shelves. Example pictures of diamonds, crystals, and rocks have been added to some of the shelves. Color the picture.

Trace each line to its matching rock. Color the rocks.

Lesson 13

Draw lines to connect the dots, starting at the number 1. Color the picture. Note: There are two separate sets of numbers.

© GOOD AND BEAUTIFUL

Draw lines to connect each rock to its shadow. Color the rocks.

Draw lines to connect the matching landforms. Color the pictures.

Draw a line to connect each of the pieces at the bottom of the page to its matching shape on the picture. Color the picture.

Extra Doodling and Drawing Page

2.0NSP460-159825 Printed in USA Jan-2025